FIBRE CHANNEL:

CONNECTION TO THE FUTURE

SECOND EDITION

The Fibre Channel Industry Association
www.fibrechannel.com

Second Edition

Copyright © 1998 by the Fibre Channel Industry Association

Printed and bound by CPI Group (UK) Ltd, Croydon, CR0 4YY
Transferred to Digital Print 2011

Cover design: Brian McMurdo, Ventana Studio, Valley Center, CA
Published by LLH Technology Publishing, Eagle Rock, VA

Library of Congress Cataloging-in-Publication Data

Fibre channel : connection to the future / the Fibre Channel
    Association. - - 2nd ed.
        p.   cm.
    Includes index.
    ISBN 1-878707-45-0 (softcover)
    1. Fibre Channel (Standard) I. Fibre Channel Association.
TK7895.B87F43   1998   98-14161
004.6'6—dc21   CIP

FOREWORD

The information explosion and the need for high-performance communications for server-to-storage and server-to-server networking have been the focus of much attention during the 90s. Performance improvements in storage, processors, and workstations, along with the move to distributed architectures such as client/server, have spawned increasingly data-intensive and high-speed networking applications. The interconnect between these systems and their input/output devices demands a new level of performance in reliability, speed, and distance.

Fibre Channel, a highly reliable, gigabit-per-second interconnect technology, allows concurrent communications among workstations, mainframes, servers, data storage systems, and other peripherals using common storage and network protocols. It provides interconnect systems for multiple topologies that can scale to a total system bandwidth on the order of a terabit per second. Fibre Channel delivers a new level of reliability and throughput. Switches, hubs, storage subsystems, storage devices, and adapters are among the products that are on the market today, providing the ability to implement a total system solution.

The members of the Fibre Channel Industry Association (FCIA) realize that wading through the entire Fibre Channel standards documentation can be a daunting task. This brief tutorial offers a readable overview of the standard. It is intended for those who are interested in a general overview of Fibre Channel and its applications. While it won't make you an expert, it provides a good starting point. We hope you find it useful.

Fibre Channel Industry Association

CONTENTS

Chapter 1

WARP SPEED INFORMATION DELIVERY

Fibre Channel is the solution for Information Technology (IT) professionals who demand reliable, cost-effective information storage and delivery at blazing speeds. Today's data explosion presents unprecedented enterprise challenges in data warehousing, imaging, integrated audio/video, networked storage, real-time computing, CAD/CAE, archival/backup, and disaster recovery. Fibre Channel is simply the most reliable, highest performance solution for information storage, transfer, and retrieval available today.

With development started in 1988 and ANSI standard approval in 1994, Fibre Channel is the mature, safe solution for gigabit communications. Fibre Channel economically and practically meets the challenge with these advantages:

- *Price Performance Leadership* – Fibre Channel delivers cost-effective solutions for storage and networks.

- *Solutions Leadership* – Fibre Channel provides versatile connectivity with scalable performance.

- *Reliable* – Fibre Channel provides sustained, assured information delivery using a reliable transport.

- *Gigabit Bandwidth Now* – Gigabit solutions are in place today! On the horizon is multiple gigabit-per-second data delivery.

- *Multiple Topologies* – Dedicated point-to-point, shared loop, and switched fabric topologies meet application requirements.

- *Multiple Protocols* – Fibre Channel delivers data. SCSI, TCP/IP, video, or raw data can all take advantage of high-performance, reliable Fibre Channel technology.

- *Scalable* – From single point-to-point gigabit links to integrated enterprises with hundreds of servers and storage systems, Fibre Channel delivers unmatched performance.

- *Congestion Free* – Fibre Channel's credit-based flow control delivers data as fast as the destination buffer is able to receive it.

- *High Efficiency* – Real price performance is directly correlated to the efficiency of the technology. Fibre Channel has very little transmission overhead, offering the capability to deliver very high, sustained utilization rates without loss of data. Most important, the Fibre Channel protocol is specifically designed for highly efficient operation using hardware.

WHY FIBRE CHANNEL?

Never before have we witnessed today's growth rate of storage capacity and distributed applications that rely so completely on high-speed communications. Visionaries in the computer industry saw this coming in the 1980s and initiated the development of a new standard to answer the data transfer requirements of this new trend. The core Fibre Channel standard, ANSI X3.230-1994, was the result.

In fact, multiple terabytes of Fibre Channel interfaced storage are installed every day. Fibre Channel works equally well for storage, networks, video, data acquisition, and many other applications. Fibre Channel is ideal for reliable, high-speed transport of digital audio/ video. Aerospace developers are using Fibre Channel for ultra-reliable, real-time networking, as well as for low-latency, high-throughput signal processing applications.

Fibre Channel is a fast, reliable data transport system that scales to meet the requirements of any enterprise. Today, installations range from small post-production systems on Fibre Channel loops to very large CAD systems linking thousands of users into a switched, Fibre Channel network.

Fibre Channel is ideal for these applications:

- High-performance storage
- Large databases and data warehouses
- Storage backup systems and recovery
- Server clusters
- Network-based storage

- High-performance workgroups
- Campus backbones
- Digital audio/video networks
- Embedded military sensor, processing, and displays
- Industrial control systems

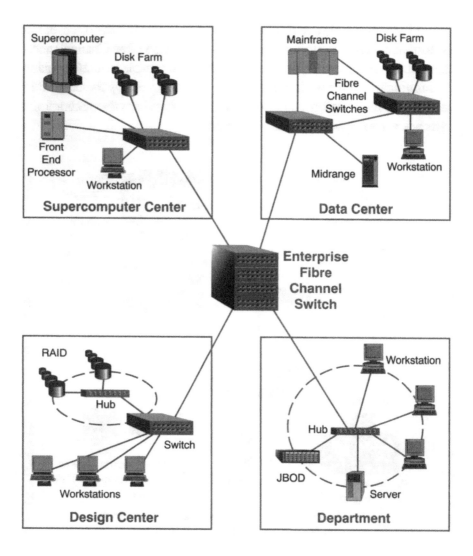

Figure 1.1 Fibre Channel meets the demands of IT systems

FIBRE CHANNEL SYSTEMS

Fibre Channel systems are assembled from adapters, hubs, storage, and switches. Host bus adapters are installed into hosts in the same manner as a SCSI host bus adapter or a network interface card. Hubs link individual elements together to form a shared bandwidth loop. Disk systems integrate a loop into a storage backplane. A port bypass circuit provides the ability to hot swap Fibre Channel disks and links to a hub. Fibre Channel switches provide scalable systems of almost any size.

IT systems today require an order of magnitude improvement in performance. High-performance, gigabit-per-second Fibre Channel meets this requirement. Fibre Channel is one of the most reliable, scalable, gigabit communications technologies today. It was designed by the computer industry for high-performance communications, and no other technology matches its total system solution capabilities.

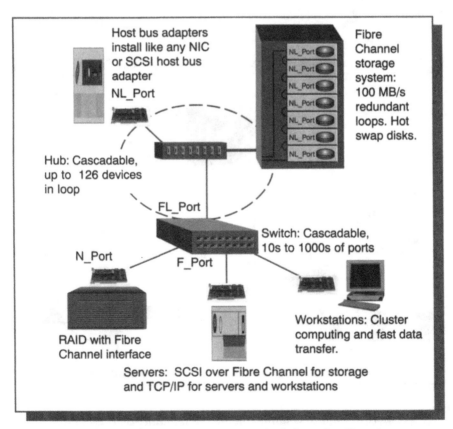

Figure 1.2 Fibre Channel systems are built from familiar elements

Chapter 2

FIBRE CHANNEL AT WORK

Visionary IT managers are delivering bottom-line improvements for their operations with Fibre Channel. They deploy Fibre Channel to take advantage of the unique combination of channel and network features. Fibre Channel networks concurrently provide networked access to storage using storage protocols and server-to-server communications using network protocols.

Virtually all data has become mission critical. Fibre Channel moves and shares this data in a timely and cost-effective manner. Fibre Channel networks are cost-effective data transport systems for many applications.

In this chapter, we show how Fibre Channel is delivering cost-effective solutions for the following applications:

- Disaster Recovery
- Multiplatform Storage
- Enterprise Decision Making
- Effective Use of IT Resources
- Explosive Data Growth
- Management
- Real-Time Applications
- Audio-Video
- Digital Imaging

Fibre Channel is the most cost-effective gigabit communications link and it will continue to be in the future. Fibre Channel's efficient structure delivers excellent price/performance when compared with other gigabit technologies. Cost can be scaled to the application.

DISASTER RECOVERY

One of today's most valuable assets is stored information that is moved, utilized, and shared in a timely and cost-effective manner. Banking records, credit card transactions, and inventory status are typical examples of critical pools of stored data.

Despite the strongest efforts, no one can avoid disasters that threaten the life of a company, like hurricanes, floods, fires, extended power outages, or sabotage. The need for backup and recovery is well documented. Never before have IT staff had the ability to deploy

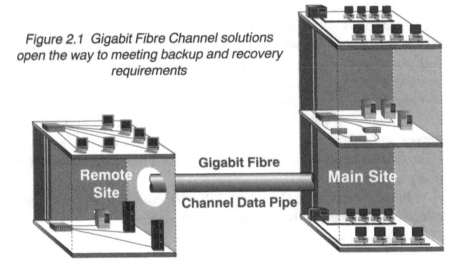

Figure 2.1 Gigabit Fibre Channel solutions open the way to meeting backup and recovery requirements

gigabit Fibre Channel solutions that enable backups and recoveries to be completed on-line or within a short window of opportunity.

Fibre Channel networks can be deployed between sites within a 10 km range, stretched to 30 km (18 miles) with optical extender modules, or, in the future, linked to the WAN using Synchronous Transfer Mode (STM) or Asynchronous Transfer Mode (ATM).

A Fibre Channel investment before a disaster strikes can dramatically impact the efficiency of a recovery. If your main IT operations were destroyed, could you quickly implement a plan to maintain day-to-day business activities? Fibre Channel will not only speed up the recovery process but also help you make an effective recovery, reducing the amount of lost productivity and revenue.

MULTIPLATFORM STORAGE

A typical company has reorganized, combined departments, decentralized, recentralized, bought other companies, outsourced IT operations, and brought IT operations back in-house. The result is an array of products that IT departments are challenged with integrating and utilizing. They must preserve this investment and still meet the information processing, storage, and distribution requirements of the enterprise.

Fibre Channel solutions offer IT managers new tools to meet this challenge, providing these benefits:

- Investment protection

- Centralized management of storage

- More efficient use of storage

- Centralized backup and recovery

- Enterprise-wide access to storage

Fibre Channel networks interface storage devices and deploy scalable gigabit storage-to-server and server-to-server communications. They may use storage and network protocols concurrently. For example, server-to-storage links using SCSI and server-to-server communications with IP create new possibilities for optimizing storage utilization. Network storage means more than NFS or FTP. It means servers sharing RAID, servers sharing tape backup systems, workstations sharing SCSI peripherals, users working on the same file, and protection of investments in SCSI-based systems.

Users who have implemented shared storage have found it helps slow down the explosive demand for storage capacity. Instead of duplicating files, they are effectively shared using Fibre Channel.

IT managers do their best to cost-effectively utilize all resources, providing reliable, responsive service to internal and external customers. Fibre Channel solutions integrate servers and storage using enterprise-wide networks unencumbered with desktop traffic. The result is better management of storage, better desktop performance, and the means to implement client/server applications effectively.

ENTERPRISE DECISION MAKING

Information-savvy organizations make data sharing and distribution a key part of their corporate structure. Unless accurate and timely data, the lifeblood of any organization, is kept flowing, decision making suffers. IT managers who use Fibre Channel-based strategies to support enterprise decision making have these advantages:

- Scalable, gigabit storage and server networking on one seamless interconnect
- Data warehouses optimized for queries
- Data marts optimized for decision analysis
- Continually expanding databases
- Expedient, low-latency access and delivery of data

The results are the ability to give decision makers high-speed access to data at any time, supporting critical business goals and objectives.

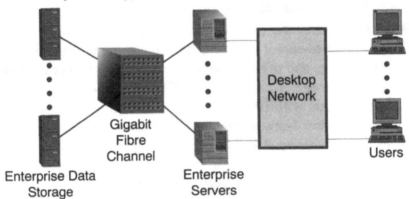

Figure 2.2 Fibre Channel solutions offer the means to reap full benefit from data

Enterprise-wide quick access and retrieval from large databases and data warehouses provide the means to solve critical business issues. When organizations gain insight to the value of Fibre Channel-enabled data access, they will be able to capture the business value inherent in expedient access to data.

Knowledge for decision making can be derived from the vast amounts of data available to businesses. Fibre Channel is here today and available to assist organizations in extracting knowledge.

EFFECTIVE USE OF IT RESOURCES

Gigabit Fibre Channel networks take advantage of previously untapped power in storage, servers and workstations. Fibre Channel networks make clusters possible with:

- Reliable communications
- Scalable networks
- Highly available systems
- Low-latency messaging
- High-bandwidth transfers

Instead of replacing systems to meet expanded requirements, Fibre Channel clusters enhance capabilities of the installed base and preserve your investment. IT managers use Fibre Channel's flexibility and reliability to add incremental processing and storage for "just-in-time computing and storage."

Fibre Channel is an open ANSI standard, connecting a heterogeneous mix of servers, workstations, and storage. Any topology is implemented as required, meeting the needs of each application.

- *High Availability* – Gigabit Fibre Channel networks enable assured communications, with servers and storage devices backing up each other. If a device fails or is taken down for maintenance, Fibre Channel connectivity enables its activity to be performed by another device without losing functional capability.

- *Compute Clusters* – Servers and workstations linked on a Fibre Channel network cooperate to process applications. Fibre Channel networks provide low-latency messaging for coordination and high-bandwidth storage interfaces. Load sharing dispatches tasks to idle units, putting unused compute cycles to work.

- *Storage Pools* – Protocols such as SCSI and IP are both effectively used to provide shared storage. Direct access to disks uses the SCSI protocol. IP-based network storage using FTP or NFS is faster using gigabit Fibre Channel.

Fibre Channel-powered clusters cost-effectively share data, share processing, put unused capacity to work, and provide high availability. The net result is more effective use of IT resources.

EXPLOSIVE DATA GROWTH

Information storage has become a critically important element for enterprises. New applications and image-oriented data have fueled a storage explosion. Storage capacity is doubling every year with no end in sight. IT managers are demanding better ways of maintaining and increasing storage performance while rapidly adding capacity to meet new demands.

Fibre Channel storage and networks deliver greater data loads, streamline data access and distribution, and handle more complex data types. IT managers can use Fibre Channel storage and networks to deliver these advantages:

- *Managed Growth* – Storage is no longer isolated behind a server. It is shared among servers, providing more efficient use of storage assets.

- *Scalable Performance* – Storage systems may now scale to tens, hundreds, or thousands of terabytes. Fibre Channel networks support this growth, scaling from tens of nodes on a shared gigabit loop to thousands of nodes on switched fabrics.

- *Accessibility* – Fibre Channel spans a building, a campus, or, with fiber optic extenders, remote sites up to 30 km away.

- *Heterogeneous Systems* – Servers, storage, and workstations from different manufacturers are interconnected with Fibre Channel networks.

- *Affordable* – Fibre Channel is the most cost-effective gigabit technology. Fibre Channel's efficient structure delivers excellent price/performance when compared with other gigabit technologies.

- *High Availability*–Fibre Channel disks come with redundant gigabit interfaces. Network architectures support high availability. Gigabit switches are now providing redundant components and automatic failover in case of failures.

- *Storage Backup* – With storage expanding and backup windows shrinking, gigabit Fibre Channel storage becomes a necessity.

As client/server systems continue to grow, enterprises need more reliable, more scalable, more intelligent storage solutions. Fibre Channel storage and networks meet these needs with an environment that consolidates and simplifies storage access.

MANAGEMENT

Many companies are returning to the concept of centralized management of data storage, even within distributed IT architectures. The Gartner Group forecasts that over 70 percent of shared storage in networked environments will be centralized by the year 2000.

Storage must be viewed as a system, delivering services and protecting data assets. Proper management of this system provides highly available data access, improved performance, complete data security, and storage growth at a reasonable cost. A storage system is comprised of on-line storage, near-line storage, archived storage, and backup storage. Storage management software moves data among these elements as required to meet the enterprise's storage management strategy. Fibre Channel removes barriers associated with the implementation of this strategy.

Fibre Channel devices use SNMP and SCSI Enclosure Services (SES) for systems management from a central location. The Fibre Channel standard supports SNMP over IP or directly over Fibre Channel natively. Fibre Channel manufacturers normally provide a point solution for SNMP that can be integrated into an enterprise management system. Fibre Channel is designed to be self-managing with most management activity focused on determining status. New WEB-based management systems are also available for system management.

SES is similar to SNMP but is designed to obtain information from storage devices that do not implement IP or SNMP. A server will act as a proxy agent for a storage system.

Centralized storage simplifies management. Fibre Channel provides the ability to view storage as if it were centralized, even if it is physically distributed around the enterprise. Fibre Channel supports critical business applications with performance, reliability, fast data access and transfer, and managed storage and server networks.

REAL-TIME APPLICATIONS

The speed and reliability of Fibre Channel is being applied to real-time applications. For example, the B-1 bomber is now using Fibre Channel for improved performance in the mission avionics suite.

Three different approaches for real-time, mission-critical applications are being developed:

- Priority/pre-emption switching fabric
- Real-time, isochronous Fibre Channel loop
- Real-time, fractional bandwidth virtual circuits

The avionics industry is moving to Fibre Channel to take advantage of the inherent benefits offered by the Fibre Channel architecture:

- Credit-based flow control
- 8B/10B serial data encoding scheme
- Bounded media access latency, providing deterministic operational behavior
- No data loss, even at 98% utilization during a streamed transfer
- No decrease in data throughput with increased link distances

Fibre Channel offers real-time applications the ability to keep pace with data and signal processors as well as new sensor technology. Industrial control systems, as well as the military, will benefit from its speed and reliability.

Figure 2.3 Fibre Channel enhances real-time applications

AUDIO-VIDEO

Audio and video are rapidly moving to an all-digital format. The Fibre Channel industry is working with the television and movie industry to provide performance-enhancing Fibre Channel solutions. New profiles are under development to map digital audio and video onto Fibre Channel. High-bandwidth, redundant Fibre Channel storage and workstation links deliver scalable, flexible, and reliable systems. Bottlenecks and single points of failure are eliminated. Both fast file transfer and audio/video stream transfers (compressed and uncompressed) are supported.

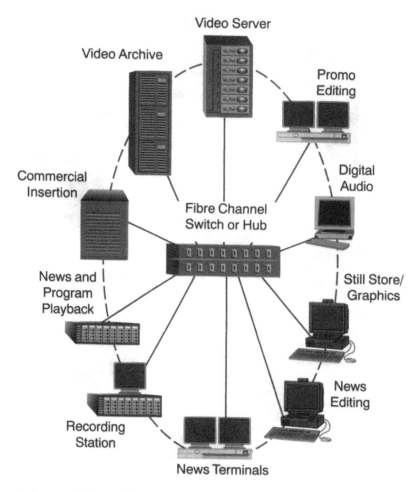

Figure 2.4 Fibre Channel makes the digital newsroom practical

DIGITAL IMAGING

Digital imaging applications, like movie post-production, pre-press, and medical imaging, typically move extremely large files and need to move these files in a very short time. Users of these files do not have the time to wait for slow delivery. The speed and reliability of Fibre Channel deliver increased productivity which, in turn, improves the bottom line. Typical results are 50 percent savings in storage requirements and 30-50 percent less time to produce projects.

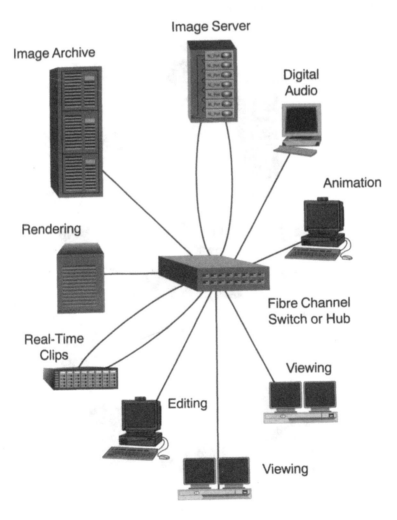

Figure 2.5 Fibre Channel improves productivity of imaging applications

Chapter 3

FIBRE CHANNEL OVERVIEW

After a lengthy review of existing equipment and standards, the Fibre Channel standards group realized that channels and networks should be able to share the same fibre. (Note that "fibre" is used throughout this book as a generic term which can indicate either an optical or a copper cable.)

IT systems frequently support two or more interfaces, and sharing a port and media makes sense. This reduces hardware costs and the size of the system, since fewer parts are needed. Fibre Channel, a family of ANSI standards, is a common, efficient transport system supporting multiple protocols or raw data using native Fibre Channel guaranteed delivery services. Profiles define interoperable attributes, based on these standards, for using Fibre Channel for different protocols or applications.

American National Standards Institute

National Committee for Information Technology Standards

Technical Committee T11: I/O Interface

▼

Fibre Channel Standards
X3.230-1994–Fibre Channel Physical and Signaling Standard (FC-PH)
Initial core standard

▶

Interoperable Profiles

Storage
Video
Networks
Real Time
Backbone
Avionics

The ambitious requirements given the standards group were:

- Performance from 133 megabits/second to four gigabits/second

- Support for distances of up to 10 km

- Small connectors

- High-bandwidth utilization with distance insensitivity

- Greater connectivity than existing multidrop channels

- Broad availability (i.e., standard components)

- Support for multiple cost/performance levels, from small systems to supercomputers

- Ability to carry interface command sets of multiple existing protocols, including Internet Protocol (IP), SCSI, IPI, HiPPI-FP, and audio/video

Fibre Channel, a channel/network standard, contains network features that provide the required connectivity, distance, and protocol multiplexing. It also supports traditional channel features for simplicity, reliability, repeatable performance, and guaranteed delivery. Fibre Channel also works as a generic transport mechanism.

Fibre Channel architecture represents a true channel/network integration with an active, intelligent interconnection among devices. All a Fibre Channel port has to do is manage a simple point-to-point connection. The transmission is isolated from the control protocol, so point-to-point links, arbitrated loops, and switched topologies are used to meet the specific needs of an application (see Chapter 6 for an explanation of topologies). The fabric is self-managing. Nodes do not need station management, which greatly simplifies implementation.

INTEROPERABILITY

Two independent laboratories do Fibre Channel testing. The Interoperability Laboratory (IOL) at the University of New Hampshire develops test suites for vendors to check compliance with the Fibre Channel standard. The Computational Science and Engineering Laboratory at the University of Minnesota is focused on functionality and extending the application of Fibre Channel.

STORAGE

Fibre Channel is the next storage interface. Fibre Channel has been adopted by the major computer systems and storage manufacturers as the next technology for enterprise storage. It eliminates distance, bandwidth, scalability, and reliability limitations of SCSI.

Storage Devices and Systems:

Fibre Channel is being provided as a standard disk interface. Industry leading RAID manufacturers are shipping Fibre Channel systems. Soon, RAID providers will not be regarded as viable vendors unless they offer Fibre Channel.

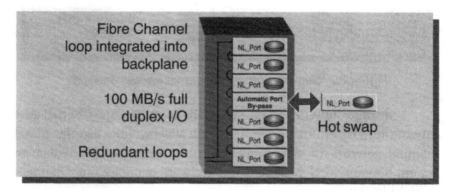

Figure 3.1 Fibre Channel disks define a new standard of performance

Storage Area Network (SAN):

The network behind the servers linking one or more servers to one or more storage systems is the SAN. Each storage system could be RAID, tape backup, tape library, CD-ROM library, or JBOD (Just a Bunch of Disks). Fibre Channel networks are robust and resilient, with these features:

- Shared storage among systems
- Independent scaling of server and storage capacities
- High performance
- Robust data integrity and reliability
- Fast data access and backup

In a Fibre Channel network, legacy storage systems are interfaced using a Fibre Channel-to-SCSI bridge. IP is used for server-to-server and client/server communications.

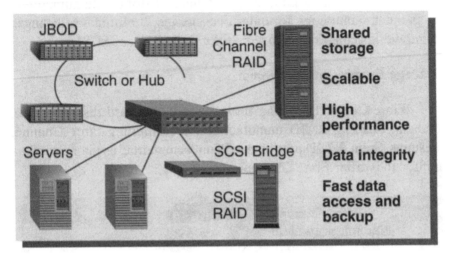

Figure 3.2 Networks behind the server deliver flexibility

Storage networks operate with both storage (SCSI) and networking (IP) protocols. Servers and workstations use the Fibre Channel network for shared access to the same storage device or system. Legacy SCSI systems are interfaced using a Fibre Channel-to-SCSI bridge.

Fibre Channel products have established a new level of performance, delivering a sustained bandwidth of over 90 MB/second for large file transfers and over ten thousand I/Os per second for business-critical database applications. This new capability for open systems storage is the reason Fibre Channel is the connectivity standard for storage access.

NETWORKS

Fibre Channel networks provide enterprises new levels of performance and reliability. The many network applications for Fibre Channel include:

- Nonstop corporate backbone

- High-performance CAD/CAE network

- Movie animation and post-production projects to reduce time to market

- Quick-response network for imaging applications

Fibre Channel was developed by the computer industry for IT applications. Its development focused on removing the performance barriers of legacy LANs. Among the performance-enhancing features of Fibre Channel for networking are:

- Confirmed delivery, enhancing the reliability of the protocol stack or the option of bypassing the protocol stack for increased performance.

- Complete support for traditional network self-discovery. Full support of ARP, RARP, and other self-discovery protocols.

- Support for dedicated bandwidth point-to-point circuits, shared bandwidth loop circuits, or scalable bandwidth switched circuits.

- True connection service or fractional bandwidth, connection-oriented virtual circuits to guarantee quality of service for critical backups or other operations.

- The option of real circuits or virtual circuits.

- Instant circuit setup time measured in microseconds using hardware enhanced Fibre Channel protocols.

- Extremely low-latency connection and connectionless service.

- Automatic self-discovery and configuration of Fibre Channel topologies.

- Full support for time synchronous applications like video, using fractional bandwidth virtual circuits.

- Efficient, high-bandwidth, low-latency transfers using variable length (0-2KB) frames. Highly effective for protocol frames of less than 100 bytes, as well as bulk data transfers using the maximum frame size.

In the early days, a single computer vendor provided a proprietary solution to a single buyer, the data processing manager. With the minicomputer the process changed, and departments bought their own computing solution. The market transitioned to multiple solutions sold to multiple buyers, resulting in incompatible, proprietary data processing systems. Over time, users realized they needed to combine all data processing into an integrated environment. This requirement opened the door for open, standards-based solutions. Now, companies are connecting their mainframes with enterprise and department servers for distributed client/server architectures.

Distributed computing and parallel processing has resulted in a significant increase in process-to-process communications. At the same time, the data storage requirements have exploded. This new paradigm only works if data can be moved and shared quickly. The need for very high-bandwidth and extremely low-latency I/O is paramount. Fibre Channel is the solution that delivers.

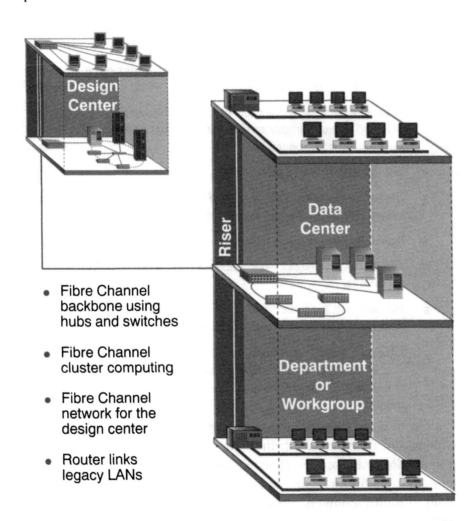

- Fibre Channel backbone using hubs and switches

- Fibre Channel cluster computing

- Fibre Channel network for the design center

- Router links legacy LANs

Figure 3.3 Fibre Channel networks deliver higher performance

Fibre Channel is attractive because it offers a standards-based solution. With the emphasis on open systems, end users are shying away from proprietary solutions and vertically integrated, single-provider solutions. Today, these users are integrating the best the industry has to offer into seamless systems.

These new systems are being driven by the technology and marketing forces associated with client/server implementations. Fibre Channel is the only technology available with the reliability, responsiveness, scalability, high throughput, and low latency needed to meet the broad range of market and technology requirements.

Users enjoy these advantages:

● Scalable systems

● More cost-effective systems

● Straightforward migration to Fibre Channel

● Continued support of legacy systems

● Graceful upward migration

Fibre Channel's scalability provides a continued return on investment long into the future.

TECHNOLOGY COMPARISON

Fibre Channel is a product of the computer industry. Fibre Channel was specifically designed to remove the barriers of performance existing in legacy LANs and channels. In addition to providing scalable gigabit technology, the architects provided flow control, self-management, and superior reliability.

Gigabit Ethernet is attractive because of its extension of LAN capabilities. It is designed to enable a common frame from the desktop to the backbone. However, Fibre Channel is designed to be a transport service independent of protocol. Fibre Channel's ability to use a single technology for storage, networks, audio/video, or to move raw data is superior to the common frame feature.

ATM was designed as a wide area network with the ability to provide Quality of Service for fractional bandwidth service. The feature of fractional bandwidth with assured Quality of Service is attractive for some applications.

For the more demanding applications, Class 4 Fibre Channel provides guaranteed delivery and gigabit bandwidth as well as fractional bandwidth Quality of Service.

Fibre Channel's use in both networks and storage provides a price savings due to economies of scale associated with larger volumes. Users can expect their most cost-effective, highest-performance solutions to be built using Fibre Channel.

As shown in Table 3.1 below, Fibre Channel is the best technology for applications that require high-bandwidth, reliable solutions that scale from small to very large.

	Fibre Channel	Gigabit Ethernet	ATM
Technology application	Storage, network, video, clusters	Network	Network, video
Topologies	Point-to-point, loop/hub, switched	Point-to-point, hub, switched	Switched
Baud rate	1.06 Gbps	1.25 Gbps	622 Mbps
Scalability to higher data rates	2.12 Gbps, 4.24 Gbps	Not Defined	2.48 Gbps
Guaranteed delivery	YES	NO	NO
Congestion data loss	Class 3 Only	YES	YES
Frame size	Variable, 0-2KB	Variable, 0-1.5KB	Fixed, 53 B
Flow control	Credit Based	Rate Based	Rate Based
Physical media	Copper and Fibre	Copper and Fibre	Copper and Fibre
Protocols supported	Network, SCSI, Video	Network	Network, Video

Table 3.1 Technology comparison

Chapter 4

PHYSICAL AND SIGNALING LAYER

Fibre Channel is structured with independent layers. The five layers of Fibre Channel, shown in Figure 4.1, define the physical media and transmission rates, encoding scheme, framing protocol and flow control, common services, and the upper-layer protocol interfaces. In this chapter, we describe the Fibre Channel Physical (FC-PH) standard, which consists of the three lower layers, FC-0, FC-1, and FC-2. The upper layers are discussed in Chapter 5.

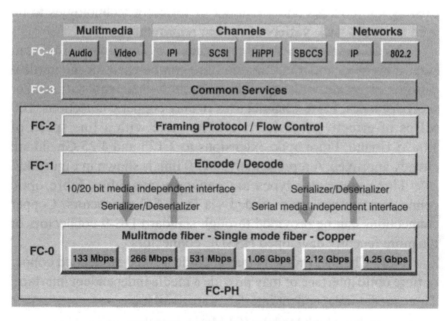

Figure 4.1 Fibre Channel layers

The three FC-PH levels are defined as follows:

- FC-0 – Covers the physical characteristics of the interface and media, including the cables, connectors, drivers (ECL, LEDs, short-wave lasers, and long-wave lasers), transmitters, and receivers.

- FC-1 – Defines the 8B/10B encoding/decoding and transmission protocol used to integrate the data with the clock information required by serial transmission techniques.

- FC-2 – Defines the rules for the signaling protocol and describes transfers of data frames, Sequences, and Exchanges.

FC-PH completed its first public review in January 1993, the second review began in October 1993, and the standard was approved in November 1994. The three levels are described in detail in the following sections.

PHYSICAL INTERFACE AND MEDIA: FC-0

The lowest level, FC-0, specifies the physical link of the channel. Fibre Channel operates over various physical media and data rates. Fibre Channel's approach ensures maximum flexibility, allowing existing cable plants and a number of different technologies to be used to meet a wide variety of system requirements.

As an example, with Fibre Channel, a single-mode fibre can enter a building and a multimode fibre can be used for the vertical distribution inside, with copper drops to individual workstations. For short distances, Fibre Channel uses twinax copper connections. 100 MBps of effective transfer rate is possible with a line speed of 1.0625 Gbaud. Fibre optic extensions to 2.125 and 4.25 Gbaud are already approved. A representative FC-0 link is shown in Figure 4.2.

Three connector types are generally available. Fibre optic connectors are usually provided via Dual SC connectors. Copper connections can be provided through standard DB-9 connectors or the more recently developed HSSDC connectors.

Fibre Channel products may include a fixed embedded copper or fibre optic interface or may provide a media independent interface. Three media independent interfaces are available.

- Gigabaud Link Modules (GLMs) include the serializer/deserializer (SERDES) function and provide a 20-bit parallel interface to the

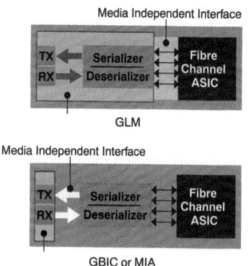

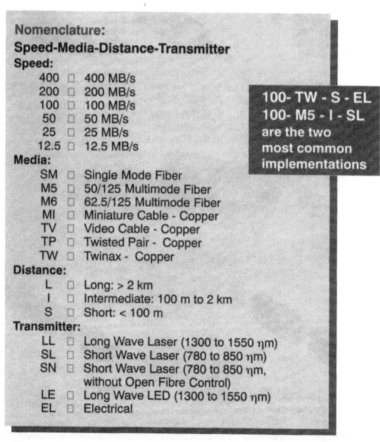

Figure 4.2 Fibre Channel physical layer provides flexibility

Fibre Channel encoding and control logic. GLMs are primarily used to provide factory configurability but may also be field exchanged or upgraded by users.

- Gigabit Interface Converters (GBICs) provide a serial interface to the SERDES function. GBICs can be hot inserted or removed from installed devices. These are particularly useful in multi-port devices such as switches and hubs, where single ports can be reconfigured without affecting other ports.

- Media interface Adapters (MIAs) allow users to convert copper DB-9 connectors to multimode fibre optics. The power to support the optical transceivers is supplied by defined pins in the DB-9 interface.

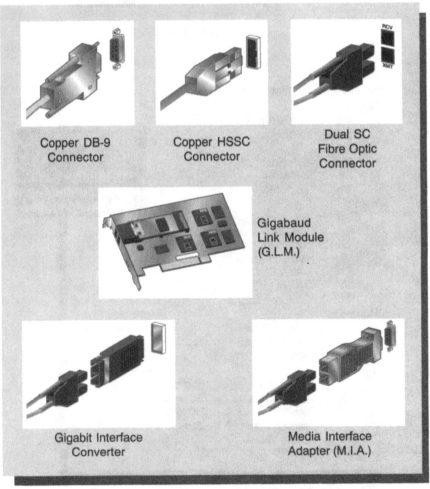

Figure 4.3 Flexible use of gigabit fibre optics or copper

TRANSMISSION PROTOCOL: FC-1

The superior transmission characteristics of a dc-balanced 8B/10B code scheme are used at FC-1 for clock recovery, byte synchronization, and encode/decode. This well-balanced code developed by IBM allows for low-cost component design and provides good transition density for easier clock recovery. A unique special character, called a comma character, ensures proper byte and word alignment. Other advantages of this code are its useful error detection capability and a simple logic implementation for both encoder and decoder. In this scheme, eight internal bits (one byte) are transmitted as a 10-bit group. The block diagram in Figure 4.4 shows the 8B/10B encoding scheme.

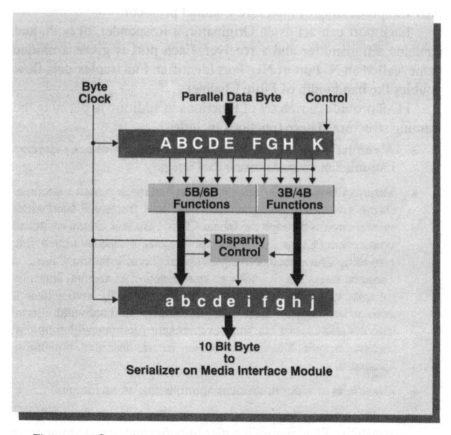

Figure 4.4 Superior transmission characteristics increase reliability

FRAMING AND SIGNALING PROTOCOL: FC-2

Reliable communications result from Fibre Channel's FC-2 Framing and Signaling protocol. FC-2 defines a data transport mechanism that is independent of upper layer protocols. FC-2 is self-configuring and supports point-to-point, arbitrated loop, and switched environments (see Figure 4.5).

An N_Port is a port on a node, such as a server, workstation, or peripheral. If a port is connected to a loop, it becomes an NL_Port. Data communications are performed over a Fibre Channel link by the interconnected ports. A node includes an ASIC with an embedded Fibre Channel Link Control Facility, which handles the logical and physical control of the link. The Link Control Facility provides a logical interface to the rest of the end system. That part of the node provides processing of the Fibre Channel protocol.

Each port can act as an Originator, a Responder, or both, and contains a transmitter and a receiver. Each port is given a unique name, called an N_Port or NL_Port Identifier. Full duplex data flow doubles the bandwidth of Fibre Channel.

FC-2 provides a rich set of functions in addition to defining the framing structure. These functions include:

- A robust 32-bit CRC (cyclic redundancy check) detects transmission errors to ensure data integrity.

- Various Classes of Service provide and manage circuit switching, frame switching, datagram services, and fractional bandwidth virtual circuits through the fabric. Class 1 is a true circuit-switched connection; Class 2 is a connectionless, frame-switched link providing guaranteed delivery and receipt confirmation; Class 3, a connectionless service with no confirmation of receipt; Intermix permits connectionless traffic simultaneously with Class 1 connection service; Class 4 supports fractional bandwidth virtual circuits; and Class 6 is a simplex connection service with multicast and preemption. The Classes of Service are described in detail in Chapter 6.

- Constructs to support efficient multiplexing of operations.

- A flow control scheme that also provides a guaranteed delivery capability. Flow control is buffer-to-buffer and/or node-to-node for connectionless service and node-to-node for connection service.

- A set of generic functions that are common across multiple upper-layer protocols.

- A built-in protocol to aid in managing the operation of the link, control the Fibre Channel configuration, perform error recovery, and recover link and port status information.

- Optional headers that may be used for network routing.

- Control information in the header to assist hardware routing.

- Process to provide segmentation and reassembly of data.

BUILDING BLOCKS

To aid in the transport of upper-layer protocol application data across the Fibre Channel network, the FC-2 layer defines a set of four "building blocks" that create logical communication from end to end.

1. Ordered Set. An ordered set consists of four 10-bit characters, a combination of data characters and special characters that are used to provide certain very low-level link functions, such as frame demarcation and signaling between two ends of a link. This signaling provides for initialization of the link after power-on and for certain basic recovery actions.

2. Frame. A frame is the smallest indivisible packet of data that is sent on the link. Frame protocol/structure is shown in Figure 4.6. Addressing is done within the frame header. Frames are not visible to the upper-layer protocols and consist of the following fields:

- Start-of-frame delimiter (an ordered set)

- Frame header (defined by FC-2)

- Optional headers (defined by FC-2)

- Variable-length payload containing upper-layer protocol user data (length 0 to maximum 2112 bytes)

- 32-bit CRC (for error detection)

- End-of-frame delimiter (an ordered set)

Each frame or group of frames can be acknowledged as part of the flow control scheme. "Busy" and "Reject" messages are also defined to provide notification of nondelivery of a frame.

3. Sequence. Fibre Channel places no limits on the size of transfers between applications (in LANs, the software is sensitive to the maximum frame or packet size that can be transmitted). Frame sizes are transparent to software using Fibre Channel because an upper-layer protocol data structure, called a Sequence, is the unit of transfer (see Chapter 5 for a discussion of the upper layers). A Sequence is composed of one or more related frames for a single operation, flowing in the same direction on the link. It is the FC-2 layer's responsibility to break a Sequence into the frame size that has been negotiated between the communicating ports and between the ports and the fabric. For example, a unit of data generated by an upper-layer protocol is disassembled into a single Sequence of one or more frames. The Sequence is also the recovery boundary in Fibre Channel. When an error is detected, Fibre Channel identifies the Sequence in error and allows that Sequence (and subsequent

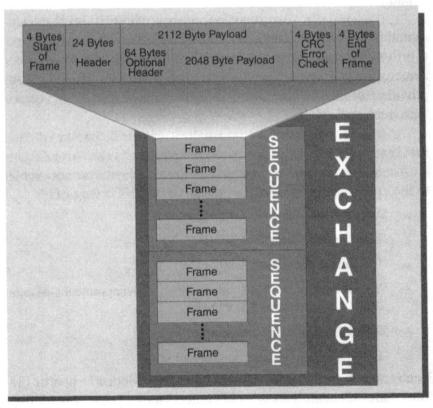

Figure 4.5 Fibre Channel Frame/Sequence/Exchange structure

Sequences) to be retransmitted. Each Sequence is uniquely identified by the initiator of the Sequence via the Sequence Identifier (SEQ_ID) field within the frame header. Additionally, each frame within the Sequence is uniquely numbered with a Sequence Count (SEQ_CNT).

4. Exchange. An Exchange is composed of one or more nonconcurrent Sequences for a single operation. For example, an operation may consist of several phases: a command to read some data, followed by the data, followed by the completion status of the operation. Each phase of command, data, and status is a separate Sequence, but they can form a single Exchange. Within a single Exchange, only a single Sequence may be flowing at any one time, although Sequences for different Exchanges may be concurrently active. This is one form of multiplexing supported by Fibre Channel. The Exchange is uniquely identified by each participating N_Port.

An Originator Exchange ID (OX_ID) is assigned, in an implementation-dependent manner, by the originating N_Port (initiator of the first Sequence), and a Responder Exchange ID (RX_ID) is assigned, in an implementation-dependent manner, by the responding N_Port (recipient of the first Sequence). The Exchange IDs are contained within the frame header and are used locally by the N_Ports to manage the Exchange.

Chapter 5

UPPER LAYERS

Fibre Channel is a transport service, moving data reliably and fast, with the scalability to meet virtually any application requirements. The Generic Services, which were defined in FC-4 along with the Upper Layer Protocol (ULP) mapping, enhance the functionality and provide common implementations for interoperability.

GENERIC SERVICES

Functions such as the Name Server and the Alias Server, which are described below, are available to all N_Ports. Additional services are described in Chapter 7.

- **Name Server** – The Name Server provides a directory of N_Ports within a fabric. The Name Server maintains a database of such items as ID, worldwide name, and supported FC-4s for each N_Port. An N_Port may query the Name Server at any time to discover other N_Ports attached to the fabric.

- **Alias Server** – The Alias Server is used to assign alias IDs to multicast groups and hunt groups. A multicast frame is delivered to all N_Ports assigned to the alias ID for the multicast group. A hunt group is a set of associated N_Ports attached to a single node. The Alias Server assigns an alias identifier that allows any frames containing this alias ID to be routed to any available (non-busy) N_Port within the set. This improves efficiency by decreasing the chance of reaching a busy N_Port.

ULP MAPPING

FC-4 defines interoperable implementations of Fibre Channel for standard protocols, audio/video, or applications like real-time computing. Each Fibre Channel implementation is specified in a separate FC-4 document.

Fibre Channel allows both network and channel protocols to be concurrently transported over the same physical interface. The following protocols are currently specified or proposed as FC-4s (proprietary protocols are also possible and permitted):

- Small Computer System Interface (SCSI)

- Internet Protocol (IP)

- Intelligent Peripheral Interface (IPI)

- High Performance Parallel Interface (HiPPI) Framing Protocol

- Link Encapsulation (FC-LE) using International Standard (IS) IS8802.2

- Single Byte Command Code Set Mapping (SBCCS) to implement ESCON® and block multiplex interfaces

- Audio Video Fast File Transfer and Real Time Stream Transfer

- Real-time, embedded avionics

FC-4 maps protocols to the FC-PH physical and signaling standard. Through mapping rules, a specific FC-4 describes how ULP processes of the same FC-4 type interoperate. A networking example of an active ULP process is the transfer of an IS8802-2 PDU from one FC node to another. A channel example of a ULP process is a SCSI operation between a channel and a disk drive.

Fibre Channel implementation is accomplished in hardware, providing faster throughput. Typically, channel protocols follow a command/data/status paradigm. Each of these information categories has different attributes and requires separate processing. However, the processing of each category is common for all protocol types. Common definitions of these information categories allow common implementations in hardware. Fibre Channel's hardware implementation eliminates compute cycles for I/O processing, making servers, workstations, and storage systems more efficient.

ESCON is a registered trademark of IBM Corporation.

Chapter 6

CLASSES OF SERVICE
AND TOPOLOGIES

Fibre Channel provides five different Classes of Service:

- Class 1 – Acknowledged Connection Service

- Class 2 – Acknowledged Connectionless Service

- Class 3 – Unacknowledged Connectionless Service

- Class 4 – Fractional Bandwidth Connection-Oriented Service

- Class 6 – Uni-Directional Connection Service

Fibre Channel connects nodes using three physical topologies that can have variants. Topologies include:

- Point-to-point

- Loop

- Switched

Loops connected to the enterprise through a switch are called public loops. An isolated loop is called a private loop. Hubs are normally used to connect nodes into a loop topology. The facility that connects multiple N_Ports is called a "fabric."

When Class of Service is combined with the various choices of physical topologies, Fibre Channel offers an unprecedented capability for delivering unmatched performance, and the ability to tune configurations to optimally meet application needs.

Switch Topology. The primary function of a switch is to receive frames from a source node and route them to the destination node. Each node has a unique Fibre Channel address, used to make the frame routing. The topology, routing path selection within a switch, and the internal structure are transparent to the nodes.

Fibre Channel switches relieve each individual port of the responsibility for station management. Each node only has to manage a simple point-to-point connection between itself and the switch.

No complex routing algorithms are used. A node simply does the equivalent of dialing a phone number by entering an identification number for the destination node in the header preceding the payload data. If it is an invalid number, the switch rejects it. If for some reason the switch cannot process the data, it responds with a "Busy" signal and the node tries again.

Think of a Fibre Channel switch as similar to the telephone system. To reach another number, you simply pick up the phone and dial. The telephone system takes responsibility for finding the number dialed and making the connection. Fibre Channel delivers the same low-cost, highly reliable, scalable connectivity and service.

Fibre Channel switches appear as single interconnecting entities to the nodes. Switch ports are the access points for physically connecting nodes. Communication between nodes and a switch or node-to-node is performed according to the Fibre Channel standard. During initialization, a node logs in, automatically discovers the type of topology to which it is connected, and operates accordingly.

A switch accepts the responsibility for routing frames according to the class of service selected. The only consideration needed is the point-to-point attachment of each node to the switch. Distance limitations are resolved by choosing the media and components needed to meet the application's requirements.

Loop and Hub Topologies. A Fibre Channel loop topology connects up to 127 ports. Loop topologies provide resilient interconnection through their use of port bypass circuits (PBCs). PBCs are installed either into the backplane of a disk enclosure or an external device called a hub. The PBC automatically detects the presence of a node and inserts it into the loop. Similarly, it detects a node that has failed or been powered off and removes it from the loop. This is accomplished automatically without manual intervention.

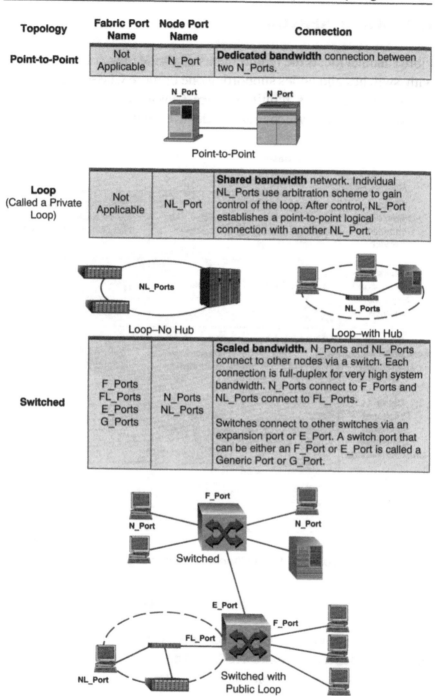

Topology	Fabric Port Name	Node Port Name	Connection
Point-to-Point	Not Applicable	N_Port	**Dedicated bandwidth** connection between two N_Ports.
Loop (Called a Private Loop)	Not Applicable	NL_Port	**Shared bandwidth** network. Individual NL_Ports use arbitration scheme to gain control of the loop. After control, NL_Port establishes a point-to-point logical connection with another NL_Port.
Switched	F_Ports FL_Ports E_Ports G_Ports	N_Ports NL_Ports	**Scaled bandwidth.** N_Ports and NL_Ports connect to other nodes via a switch. Each connection is full-duplex for very high system bandwidth. N_Ports connect to F_Ports and NL_Ports connect to FL_Ports.\n\nSwitches connect to other switches via an expansion port or E_Port. A switch port that can be either an F_Port or E_Port is called a Generic Port or G_Port.

Figure 6.1 Fibre Channel topologies

CLASSES OF SERVICE

Fibre Channel meets the requirements of a wide range of applications with switches and nodes supporting one or more Classes of Service. No manual set up is required, since Classes of Service that are supported between switches and nodes are determined during login.

	Credit-Based Flow Control	Delivery	Use
-Class 1- Acknowledged Connection Service	Node-to-Node after Connection	Reliable and Dedicated	Large Files and Full-Bandwidth Service
-Class 2- Acknowledged Connection Service	Buffer-to-Buffer Node-to-Node	Reliable and Multiplexed	Clustering, Networking, OLTP
-Class 3- Unacknowledged Connection Service	Buffer-to-Buffer	Datagram and Multiplexed	Storage, Networking, Broadcast, Multicast
-Class 4- Fractional Bandwidth Connection-Oriented Service	Buffer-to-Buffer Node-to-Node	Reliable and Guaranteed with Quality of Service	Real-Time Systems, Real-Time Audio/Video
-Class 6- Uni-Directional Connection Service	Node-to-Node after Connection	Reliable and Dedicated, assisted by Multicast Server	Video Distribution (One to Many) Data Acquisition

Table 6.1 Fibre Channel Classes of Service

Class 1: Acknowledged Connection Service

Class 1 provides true connection service. The result is circuit-switched, dedicated bandwidth connections. Fibre Channel's advantage is connection setup and tear-down measured in microseconds.

An end-to-end path between the communicating devices is established through the switch. Since the only overhead for Class 1 is connection setup and tear-down, it is an efficient service for large data exchanges. Fibre Channel Class 1 service provides an acknowledgment of receipt for guaranteed delivery. Class 1 also provides full-bandwidth, guaranteed delivery and bandwidth for applications like image transfer and storage backup and recovery. Some applications use the guaranteed delivery feature to move data reliably and quickly without the overhead of a network protocol stack.

"Camp on" is a Class 1 feature that enables a switch to monitor a busy port and queue that port for the next connection. As soon as the port is free, the switch makes the connection. This switch service speeds connect time, rather than sending a "Busy" back to the originating N_Port and requiring the N_Port to retry to make the connection.

"Stacked Connect" is a Class 1 feature that enables an originating N_Port to queue sequential connection requests with the switch. Again, this feature reduces overhead and makes the switch service more efficient.

Class 1 has a variation called buffered service. It is used to connect two Fibre Channel ports that are operating at different speeds.

Another form of Class 1 is called Dedicated Simplex Service. Normally, Class 1 connections are bidirectional; however, in this service, communication is in one direction only. It is used to separate transmit and receive switching. It permits one node to transfer to another node while simultaneously receiving from a third node.

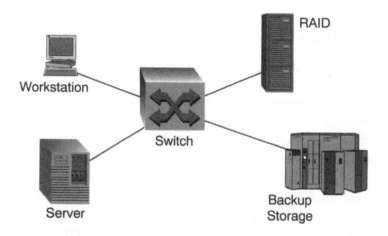

Figure 6.2 Class 1 service provides dedicated bandwidth

Class 2: Acknowledged Connectionless Service

Class 2 is a connectionless service, independently switching each frame and providing guaranteed delivery with an acknowledgment of receipt. The path between two interconnected devices is not dedicated. The switch multiplexes traffic from N_Ports and NL_Ports without dedicating a path through the switch.

Class 2 credit-based flow control eliminates congestion that is found in many connectionless networks. If the destination port is congested, a "Busy" is sent to the originating N_Port. The N_Port will then re-send the message. This way, no data is arbitrarily discarded just because the switch is busy at the time.

The very low frame latency in Class 2 makes it ideal for shorter data transfers like those in most business applications.

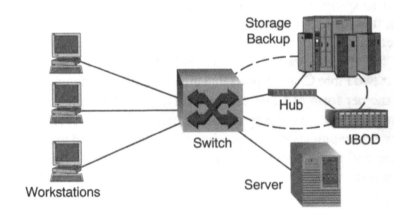

Figure 6.3 Class 2 service provides low-latency guaranteed delivery

Class 3: Unacknowledged Connectionless Service

Class 3 is a connectionless service, similar to Class 2, but no confirmation of receipt is given. This unacknowledged transfer is used for multicasts and broadcasts on networks and for storage interface on Fibre Channel loops. The loop establishes a logical point-to-point connection and reliably moves data to and from storage.

Class 3 Arbitrated Loop transfers are also used for IP networks. Some applications use logical point-to-point connections without using a network layer protocol, taking advantage of Fibre Channel's reliable data delivery.

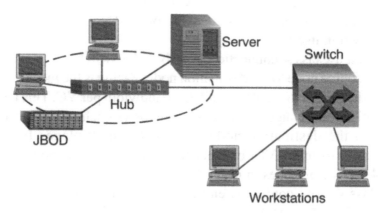

Figure 6.4 Class 3 service supports storage and networks

Class 4: Fractional Bandwidth Acknowledged Connection-Oriented Service

Class 4 is a fractional bandwidth, connection-oriented service. Virtual connections are established with bandwidth reservation for a predictable quality of service (QoS). A Class 4 connection is bidirectional, with one virtual circuit (VC) operational in each direction, and supports a different set of QoS parameters for each VC. These QoS parameters include guaranteed bandwidth and bounded end-to-end delay. A QoS facilitator (QoSF) function is provided within the switch to manage and maintain the negotiated QoS on each VC.

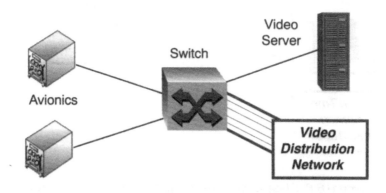

Figure 6.5 Class 4 supports real-time applications

A node may reserve up to 256 concurrent Class 4 connections. Separate functions of Class 4 are setup of the QoS parameters and the connection itself.

When a Class 4 connection is active, the switch paces frames from the source node to the destination node. Pacing is the mechanism used by the switch to regulate available bandwidth per VC. This level of control permits congestion management for a switch and guarantees access to the destination node. The switch multiplexes frames belonging to different VCs between the same or different node pairs. Class 4 service provides in-order delivery of frames.

Class 4 flow control is end-to-end and provides guaranteed delivery. Class 4 is ideal for time-critical and real-time applications like video.

Class 6: Uni-Directional Connection Service with Multicast and Preemption

Class 6 is similar to Class 1, providing uni-directional connection service. However, Class 6 also provides reliable multicast and

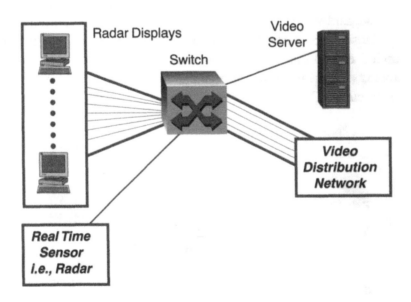

Figure 6.6 Class 6 provides dedicated one-to-many service

preemption. Class 6 is ideal for video broadcast applications and real-time systems that move large quantities of data.

Intermix

Fibre Channel has an optional mode called Intermix. Intermix allows the reservation of full Fibre Channel bandwidth for a dedicated (Class 1) connection. It also allows connectionless traffic within the switch to share the link during idle Class 1 transmissions. An ideal application for Intermix is linking multiple large file transfers during system backup. During a Class 1 file transfer, a Class 2 or 3 message can be sent to the server to set up the next transfer. Upon completion of one transfer, the next will immediately start, increasing efficiency.

Hunt Groups

A hunt group is a set of associated N_Ports attached to a single node. This set is assigned an alias identifier that allows any frames containing this alias to be routed to any available (i.e., non-busy) N_Port within the set. This improves efficiency by decreasing the chance of reaching a busy N_Port. A simple application can improve server response with multiple host bus adapters, giving the server multiple nodes in the Fibre Channel network. The net result is better utilization of the server and faster delivery to users.

Multicast

Multicast delivers a single transmission to multiple destination N_Ports. This includes sending to all N_Ports on a fabric (broadcast) or to only a subset of the N_Ports on a fabric (multicast). This provides the ability to implement effective workgroup segmentation on high-performance Fibre Channel networks.

TOPOLOGIES

Fibre Channel nodes login with each other and the switch to exchange operating information on attributes and characteristics. This information is used to establish interoperability parameters.

Whether it is a switch, an active hub, or a loop is irrelevant, because station management issues related to topology are not handled by Fibre Channel ports but are the responsibility of the switch.

Selecting the topology for an enterprise or application is determined by the system performance requirements, packaging options, and the user's growth requirements.

Fibre Channel variations in topology, Classes of Service, and physical layers offer an unprecedented capability for delivering unmatched performance, matching implementation to application needs.

The key point is that one N_Port / NL_Port design is compatible with all switch implementations. As discussed earlier, Fibre Channel matches the application requirements with multiple topologies:

- Point-to-point, dedicated bandwidth

- Loop shared bandwidth

- Switched scaled bandwidth

POINT-TO-POINT, DEDICATED-BANDWIDTH TOPOLOGY

Point-to-point topology is available in two varieties. The first is simply two N_Ports connected and exchanging information. The second (shown in Figure 6.7) is when an externally controlled link switch is used to set up different point-to-point configurations. In both cases,

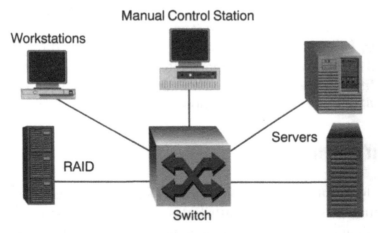

Figure 6.7 Link switches deliver low-cost reconfigurable systems

the two N_Ports have 100% utilization of the bandwidth. The communication is full-duplex; therefore, a one-gigabit transmit and receive link is delivering two Gbps of dedicated bandwidth.

LOOP SHARED-BANDWIDTH TOPOLOGY

Fibre Channel loop provides a low-cost solution for allowing multiple devices to share gigabit bandwidth. Nodes share access to the loop but have full use of the gigabit bandwidth when making logical point-to-point connections. Up to 127 nodes may be in a loop. Only one of these nodes may be a switch port. Loops are used for networks and are the replacement for SCSI as the next standard disk interface.

Nodes request control of the loop by sending a Fibre Channel signal called a primitive. If the signal is returned with the sending node's address, that node or switch port has control of the loop. If two or more nodes or the switch port are contending for control at the same time, loop control is given to the lowest address. After a node or switch port has control, it opens a full-duplex, point-to-point circuit

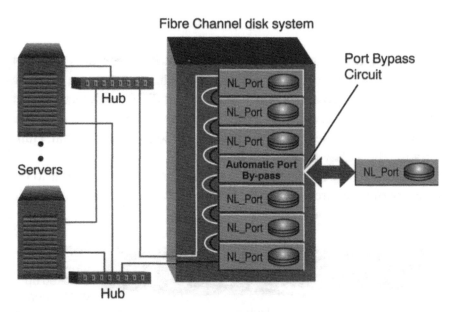

Figure 6.8 Powerful, redundant 100 MBps loops connect multiple servers with Fibre Channel storage

with a second node or switch port. Only one pair may communicate on the loop at one time. Any class of service may be used; however, most loop applications use Class 3. When control is released, another arbitration takes place. Fairness is provided by guaranteeing equal access to all ports.

The loop is self-configuring and may operate with or without a switch present. A loop node or switch port self-discovers its environment and works properly (without manual intervention) with other nodes and switches in the Fibre Channel system.

An isolated loop not connected to a switch is called a Private Loop. If the loop is linked into a Fibre Channel switch via a port called an FL_Port, the loop is called a Public Loop.

Nodes automatically relay frames unless that node has established a temporary connection with another node. Loop nodes monitor the inbound line for primitive signals addressed to that node. After a primitive signal is received directing a temporary point-to-point connection with another node, all frames received are intended for that node and are directed internal to the node. This mechanism provides a very simple and effective means of reliable loop communications.

Hubs

Loops can be wired node to node, but if a node fails or is not powered on, the loop is out of operation. This is overcome with the use of a hub. A hub uses a port-bypass circuit (PBC) to detect whether an active node is present. The PBC either opens the loop to insert the active node or closes the loop, ensuring the loop is operational. Hubs provide the ability to hot-plug nodes in and out of the loop.

Hubs provide a physical star wiring environment similar to the concept of structured wiring for LANs. Hubs can be stacked to build loops of up to 127 nodes.

Copper or optical links. Stacked up to 127 ports.

Hub

Figure 6.9 Fibre Channel hubs provide reliable loop connectivity

Fibre Channel storage enclosures have a loop built into the backplane. Each backplane node also has a PBC just like those used in hubs. Disks can be hot swapped in and out of the loop. This hot-swap feature is the foundation for Fibre Channel's 7 x 24 on-line storage management capabilities.

SWITCHED SCALED-BANDWIDTH TOPOLOGY

Fibre Channel switches are extremely easy to install and use because the Fibre Channel protocol provides self-configuration and self-management. When a Fibre Channel switch or node powers on, it determines what is on the other end of the cable and figures out how to work with it. If it is node-to-node, they automatically operate point-to-point. If it is switch-to-node, the node logs in with the switch and exchanges interoperability parameters. If it is switch-to-switch, the login process determines configuration and addresses. All of this is automatic, without operator intervention.

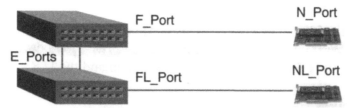

Figure 6.10 Switch ports are automatically configured

As shown in Figure 6.10, ports on a switch are either F_Ports / FL_Ports or E_Ports (expansion ports used to interconnect switch chassis). Switch ports connect to N_Ports, NL_Ports, or other switches. If a port is both an F_Port and an E_Port, it is called a G_Port (generic). Most switches come with G_Ports that automatically determine the correct use of either F_Port or an E_Port functionality during initialization and operate accordingly.

Fibre Channel switches that offer connection and connectionless switching actually are two separate switches in one. As shown in Figure 6.11, Class 1 and 6 services are provided with the circuit switch and Class 2, 3, and 4 services are provided using the frame switch. This unique Fibre Channel feature lets applications take advantage of the best features of both switching technologies. Using both a circuit

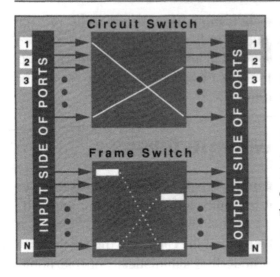

Figure 6.11 Fibre Channel switches use circuit and frame switches

and a frame switch is not required. Some switches are dedicated circuit switches and some are dedicated frame switches.

In the frame-switching mode, the bandwidth is dynamically allocated on a link-by-link basis. Based on adaptive routing within the switch, individual frames between switch ports are independently switched. During frame switching, buffering is required within the switch to provide link level flow control between the switch and the connected N_Port, NL_Port, or E_Port. Applications that need low latency for short transfers will utilize frame switching.

In the future, E_Ports will also be connected to a WAN through an Inter Working Unit (IWU). This is the result of a joint development between the Fibre Channel industry in Japan and the U.S.

Installation

During initialization, address assignment and inter-switch paths are automatically determined. At any time, nodes can be added or deleted. When nodes are connected, they automatically login and exchange operating parameters with the switch. During login, N_Ports inherit the addresses of the corresponding switch ports. NL_Ports inherit the top 16 bits of the switch FL_Port and assign up to 126 addresses using the bottom 8 bits for members of the loop.

SUMMARY

Point-to-point, loop, and switched Fibre Channel systems are deployed in mission-critical applications, delivering reliable 7x24 operations with hundreds of gigabits per second of system bandwidth.

Chapter 7

FIBRE CHANNEL SERVICES

Fibre Channel services provide a set of functions, some of which are required by Fibre Channel protocols and some of which provide optional enhancements to Fibre Channel's basic protocols. Services are typically found in switched topologies, and their functions are outlined in Table 7.1.

Login Server (Mandatory)	• Helps discover operating characteristics • Provides N_Port address assignment
Fabric/Switch Controller (Mandatory)	• Assists initialization configuration • Routing management • Optional fabric-assisted services
Name Server (Optional)	• Translation for N_Port IDs - IEEE addresses (worldwide unique) - IP addresses - Symbolic names for vital product data
Management Server (Optional)	• Configuration management • Performance management • Fault management • User diagnostics • Accounting
Time Server (Optional)	• Management of expiration timers • Time synchronization
Alias Server (Optional)	• Hunt group management • Multicast group management
Quality of Service Server (Optional)	• Class 4 Quality of Service management
Class 6 Multicast Server (Optional)	• Class 6 multicast group management

Table 7.1 Fibre Channel server functions

Fibre Channel station management functions are handled by special servers within the switch and do not have to be handled at each node. This results in a simpler, less costly system.

Multiple switches linked together form a fabric. The server functions may be centralized on a single switch in a fabric or distributed over all the switches in the fabric.

Each server is associated with a well-known, fabric-wide, unique address reserved by Fibre Channel. While the servers appear as N_Port addresses, they may be implemented internal to a fabric. The protocol for all services is the Fibre Channel common transport protocol (FC-CT), which is transparent to the fabric type or topology. The servers support all classes of service.

Login Server

The login server is a logical entity within the switch which receives and responds to the switch login requests. The login server also confirms or assigns the N_Port address of the node that initiates the login. The login server can be a proxy agent to provide a name server with the login attributes of the N_Port.

Fabric/Switch Controller

A fabric/switch controller provides both required and optional services, such as initialization, configuration, routing management, and optional fabric-assisted services.

Name Server

The primary function of the name server is to give all nodes attached to a fabric the ability to discover other nodes, their attributes, and N_Port identifiers. A node name is an 8-byte value that may be an IEEE address (worldwide unique) or may be locally assigned. In either case, it is unique within the switch address space. N_Port identifiers are assigned (optionally by the login server) when login takes place. After this point, a node or N_Port may register the association between the node and related N_Ports. Translation for IP address, upper-layer protocols, and classes of service supported is also provided.

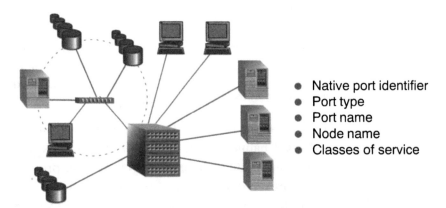

● Native port identifier
● Port type
● Port name
● Node name
● Classes of service

Figure 7.1 Name servers are a directory for Fibre Channel networks

Management Server

Fibre Channel has adopted the industry-accepted Simple Network Management Protocol (SNMP). Management Information Base (MIB) data is associated with each Fibre Channel node, hub, and fabric. SNMP is used to monitor/modify the MIB data. SNMP may use IP over FC-PH as its transport mechanism (the conventional use of SNMP). A generic solution is provided in which SNMP natively uses FC-CT over FC-PH without the requirement of IP. This allows consolidated management of Fibre Channel networks consisting of both computing equipment and peripheral devices.

Time Server

The time server manages the timers used within the Fibre Channel system, time synchronization of all the switch elements, and time stamping of messages.

Alias Server

The alias server is used to establish hunt groups and multicast groups. For example, multiple server connections to a Fibre Channel fabric with a single address could be a hunt group. The Fibre Channel switch finds the first free switch port when linking other servers or clients with the server. This is similar to finding the "first available agent" when you place a telephone call. The result is improved server utilization and better response to clients.

The alias server manages the registration and de-registration of aliases. Routing of the frames is the responsibility of the fabric.

A multicast group is a group of N_Ports that receive frames transmitted by an N_Port belonging to the same group. The frame from a source N_Port is replicated to all the other member N_Ports of the multicast group using Class 3 service. Each multicast group has its own alias identification (MG_ID). A single N_Port can register with more than one multicast group at any time. An N_Port may also register a list of N_Ports as members of one or more multicast groups.

A hunt group is a group of N_Ports under a single controlling entity. Each hunt group is associated with an alias identification (HG_ID). A hunt group provides a single alias address for all members of the group. The fabric assigns the first available N_Port within this hunt group as the destination for the service class. In Class 1, hunt groups facilitate selection of an available N_Port for a dedicated connection. In Class 2 or 3, hunt groups can be deployed to increase bandwidth, reduce latency, or both.

Quality of Service Server

The quality of service server is used to enter the desired QoS established for Class 4 service on virtual channels. The QoS server is an entity within the fabric, required for Class 4, for establishing and maintaining the QoS parameters for all dormant and active Virtual Circuits (VCs). The QoS parameters are guaranteed minimum bandwidth and guaranteed maximum end-to-end delay on each VC. The QoS server is also responsible for the management of Class 4 circuit connection set up and tear down. It makes sure that the requested QoS is provided, based on the available resources, before granting connections. All QoS parameters are negotiable for each virtual circuit.

Class 6 Multicast Server

The Class 6 multicast server is an optional fabric entity that manages the responses coming from all destination N_Ports belonging to a Class 6 multicast group. The server returns a single response to the source N_Port. Class 6 multicast provides a reliable multicast service based on dedicated connections. The frame to be multicast is sent to the Class 6 multicast server for replication and transmission to the destination N_Ports on separate dedicated connections through the fabric. The fabric is responsible for the delivery of these frames to the destination N_Ports. The frame header contains all the N_Port IDs of the multicast group used by the Class 6 multicast server. An N_Port becomes a member of the multicast group when a connection is established. When the multicast connection is removed, the N_Port ceases to be a member of that multicast group.

SUMMARY

Fibre Channel brings to market reliable, high-performance, easy-to-use, low-cost communications required by the new breed of data- and communications-intensive applications. It provides new levels of performance for storage and server networks. Fibre Channel's high-speed links offer more cost-effective solutions than today's systems.

Fibre Channel enables heterogeneous clusters of storage, servers, and workstations. It combines the attributes of a channel with those of a network, a synthesis which results in a more reliable, faster, lower-cost, simpler, and more efficient solution for IT systems.

Massive data warehouses and data marts enable better management and delivery of data for improved decision making. Highly available, scalable computer and storage clusters deliver effective use of corporate resources. Enterprises have seen a storage explosion, and Fibre Channel provides the means to manage storage growth, while maintaining and increasing performance.

Fibre Channel is being used to provide fast, reliable networks for mission avionics in military aircraft. The movie and video post-production industry achieves tremendous increases in productivity with Fibre Channel. The audio/video industry and the Fibre Channel industry are defining the Fibre Channel enabled digital studio.

Fibre Channel is a fast, reliable data transport system that scales to meet the requirements of any enterprise. Today, installations range from small post-production systems on Fibre Channel loops to very large CAD systems linking thousands of users into a switched Fibre Channel network.

Fibre Channel is truly "**cost effective.**"

SPECIAL WORKING INTEREST GROUPS (SWIGs)

The Fibre Channel Industry Association (FCIA) is organized to support the deployment of Fibre Channel applications. It has organized market segment groups responsible for ensuring that Fibre Channel meets the specific needs of these markets. These FCIA organizations, called Special Working Interest Groups or SWIGs, publish profiles that map Fibre Channel into interoperable solutions. Representative SWIGs include storage, video, avionics, real-time, and interoperability.

FOR MORE INFORMATION

The Fibre Channel Industry Association welcomes your inquiry. Please contact us at:

Fibre Channel Industry Association
404 Balboa Street
San Francisco, CA 94118
USA

Voice: (415) 750-8355
Fax: (415) 751-4829
E-mail: Info@FibreChannel.com
Web site: http://www.fibrechannel.com

For information on Fibre Channel loop activities contact:
Fibre Channel Loop Community
P.O. Box 2161
Saratoga, CA 95070
(408) 867-1385
Web site: http://www.fcloop.org

For ANSI documentation information, please contact:

Global Engineering
15 Inverness Way East
Englewood, CO 80112-5704
Phone: (800) 854-7179 or (303) 792-2181
Fax: (303) 792-2192

For the latest information on Fibre Channel standards activities:

Web site: http://www.dpt.com/t11/pub/fc

GLOSSARY

ANSI	American National Standards Institute, the coordinating organization for voluntary standards in the United States.
Arbitration	The process of selecting one respondent from a collection of several candidates that request service concurrently.
ATM	Asynchronous Transfer Mode. A type of packet switching that transmits fixed-length units of data.
Broadcast	Sending a transmission to all N_Ports on a fabric.
Channel	A point-to-point link, the main task of which is to transport data from one point to another.
Controller	A computer module that interprets signals between a host and a peripheral device. The controller often is part of the peripheral device.
CRC	Cyclic Redundancy Check, an error-correcting code used in Fibre Channel.
Datagram	Refers to the Class 3 Fibre Channel service that allows data to be sent rapidly to multiple devices attached to the fabric, with no confirmation of receipt.
8B/10B	A data encoding scheme developed by IBM, translating byte-wide data to an encoded 10-bit format.
E_Port	An expansion port on a switch. It is used to link multiple switches together into a Fibre Channel fabric.

ESCON	Enterprise Systems Connection
Exchange	A term that refers to one of the Fibre Channel "building blocks," composed of one or more nonconcurrent Sequences.
Fabric	The facility that connects multiple N_Ports.
FC-PH	Fibre Channel Physical standard, consisting of the three lower layers, FC-0, FC-1, and FC-2.
FC-0	Lowest layer of the FC-PH standard, covering the physical characteristics of the interface and media.
FC-1	Middle layer of the FC-PH standard, defining the 8B/10B encoding/decoding and transmission protocol.
FC-2	Highest layer of FC-PH, defining the rules for signaling protocol and describing transfer of frames, Sequences, and Exchanges.
FC-3	The hierarchical level in the Fibre Channel standard that provides common services.
FC-4	The hierarchical level in the Fibre Channel standard that specifies the mapping of upper-layer protocols (ULPs) to levels below.
FDDI	Fiber Distributed Data Interface. ANSI's architecture for a Metropolitan Area Network (MAN); a network based on the use of optical-fiber cable to transmit data at 100 Mbit/sec.
F_Port	Fabric port, the access point of the fabric for physically connecting the N_Port.
FL_Port	Fabric port, the access point of the fabric for physically connecting an Arbitrated Loop of NL_Ports.
Frame	An indivisible unit of information used by FC-2.
Gigabit	One billion bits, or one thousand megabits.
G_Port	Generic switch port that can be either an F_Port or an E_Port. Port function is automatically determined during login.

HiPPI	High Performance Parallel Interface, an 800- or 1600-Mbit/second interface to supercomputer networks (formerly known as high-speed channel). Developed by ANSI.
Hunt Group	A set of associated N_Ports attached to a single node and assigned a special identifier that allows any frames containing this identifier to be routed to any available N_Port within the set.
Hub	A Fibre Channel device that connects nodes into a logical loop by using an apparent star topology. Hubs will automatically recognize an active node and insert the node into the loop. A node that fails or is powered off is automatically removed from the loop.
Intermix	A mode of service defined by Fibre Channel that reserves the full Fibre Channel bandwidth for a dedicated (Class 1) connection, but that also allows connectionless traffic to share the link if the bandwidth is available.
I/O	Input/output
IP	Internet Protocol
IPI	Intelligent Peripheral Interface
L_Port	A Fibre Channel port which supports the arbitrated loop topology.
LAN	See Local Area Network
Local Area Network (LAN)	A communications system whose dimensions typically are less than 5 kilometers. Transmissions within a LAN are mostly digital, carrying data among stations at rates usually above one megabit/second.
Login Server	Entity within the Fibre Channel fabric that receives and responds to login requests.
Multicast	Refers to delivering a single transmission to multiple destination N_Ports.
Name Server	Provides translation from a given node name to one or more associated N_Port identifiers or their attributes.

N_Port	"Node" port, a Fibre-Channel-defined hardware entity at the end of a link.
NL_Port	"Node" port connected to an Arbitrated Loop.
Network	An aggregation of interconnected nodes, including workstations, file servers, and/or peripherals, with its own protocol that supports interaction.
Operation	A term, defined in FC-2, that refers to one of the Fibre Channel "building blocks" composed of one or more, possibly concurrent, Exchanges.
Ordered Set	A Fibre Channel term referring to four 10-bit characters (a combination of data and special characters) providing low-level link functions, such as frame demarcation and signaling between two ends of a link. It provides for initialization of the link after power-on and for some basic recovery actions.
Originator	A Fibre Channel term referring to the initiating device.
Port	The hardware entity within a node that performs data communications over the Fibre Channel link.
Port Bypass Circuit (PBC)	A circuit used in hubs and disk enclosures to open or close the loop automatically to add or remove nodes on the loop.
Protocol	A data transmission convention encompassing timing, control, formatting, and data representation.
Receiver	A terminal device that includes a detector and signal processing electronics.
Responder	A Fibre Channel term referring to the answering device.
SCSI	Small Computer System Interface
Sequence	A term referring to one of the Fibre Channel "building blocks," made up of one or more related frames for a single Exchange.

SONET	Synchronous Optical Network. A standard for optical network elements. Basic level is 51.840 megabit/second (OC-1); higher levels are *n* times the basic rate (OC-*n*).
Star	The physical configuration used with hubs in which each user is connected by communication links radiating out of a central hub that handles all communications.
Striping	A method for achieving higher bandwidth using multiple N_Ports in parallel to transmit a single information unit across multiple levels.
T11	A technical committee of the National Committee for Information Technology Standards, titled T11 I/O Interfaces. It is tasked with developing standards for moving data in and out of central computers.
Time Server	A Fibre Channel-defined service function that allows for the management of all timers used within a Fibre Channel system.
Topology	The logical and/or physical arrangement of stations on a network.
Transmitter	A device that includes a source of driving elements.
ULP	Upper layer protocol
Virtual circuit	A unidirectional path between two communicating N_Ports that permits fractional bandwidth.

Printed and bound by CPI Group (UK) Ltd, Croydon, CR0 4YY

03/10/2024

01040847-0011